Anne Jenisca Hilsenah
Daurelis Bothel

Access to agricultural and livestock land in Diego-Suarez

Anne Jenisca Hilsenah
Daurelis Bothel

Access to agricultural and livestock land in Diego-Suarez

Administrative Procedures and Requirements

ScienciaScripts

Imprint
Any brand names and product names mentioned in this book are subject to trademark, brand or patent protection and are trademarks or registered trademarks of their respective holders. The use of brand names, product names, common names, trade names, product descriptions etc. even without a particular marking in this work is in no way to be construed to mean that such names may be regarded as unrestricted in respect of trademark and brand protection legislation and could thus be used by anyone.

Cover image: www.ingimage.com

This book is a translation from the original published under ISBN 978-620-3-41670-1.

Publisher:
Sciencia Scripts
is a trademark of
Dodo Books Indian Ocean Ltd. and OmniScriptum S.R.L publishing group

120 High Road, East Finchley, London, N2 9ED, United Kingdom
Str. Armeneasca 28/1, office 1, Chisinau MD-2012, Republic of Moldova, Europe
Managing Directors: Ieva Konstantinova, Victoria Ursu
info@omniscriptum.com

Printed at: see last page
ISBN: 978-620-8-56102-4

Title : Access to agricultural and livestock land in Diego-Suarez

Subtitle: Administrative Procedures and Conditions

Respect

HILSENAH Anne Jenisca, BOTHEL Daurelis

SUMMARY

Land ownership is an essential pillar of development. In Madagascar, land is a symbol of wealth and security, making land control a major issue. Access to land, crucial to survival and well-being, remains a challenge, particularly in strategic areas such as Diégo-Suarez. This work aims to regulate the use of land, particularly private land, to balance private, public and environmental interests. Interviews with local authorities, notably the Commune Urbaine de Diégo-Suarez and the DIANA region's land use planning office, enabled us to analyze the conditions of access to land. The aim was to draw up an inventory of administrative procedures for the acquisition of private and public State-owned estates, large tracts of land (over 250 ha), as well as specific procedures for mining or construction activities.

Key words: Domaine Privé de l'Etat, Domaine Public, land access, administrative procedure, Diégo-Suarez

ACKNOWLEDGEMENTS

We would like to express our gratitude to the Diego-Suarez Regional Office of Land Management and to the teaching staff of the Institut Universitaire des Sciences de l'Environnement et de la Société (IUSES) at the University of Antsiranana. We would also like to thank the following people:

- Mrs. RABEMIARISOA Olivia Dany, Inspectrice des Domaines, for her help and invaluable advice;
- Monsieur le Directeur de la Commune urbaine de Diego-Suarez, for his support and collaboration during this study.

CV

- HILSENAH Anne Jenisca, Student at the Institut Universitaire des Sciences de l'Environnement et de la Société (IUSES), **major:** Sciences du Vivant de la Terre et de la Société (SVTS), **specialization:** Sciences Humains et de la Terre (SHT), University of Antsiranana (Madagascar)

-BOTHEL Daurelis, Doctorate in Science and Technology, **specializing in** biodiversity and environmental conservation, Université d'Antsiranana (Madagascar)

Table of contents

1. INTRODUCTION

On a global scale, an estimated 5 billion hectares of land are used for agriculture, representing 38% of the world's land surface. Of this land, around 1.5 billion hectares are potentially available. However, their development is limited by environmental, economic and political factors. Livestock farming, which mainly occupies uncultivable land such as grasslands, mountains, steppes and savannahs, represents around 3.5 billion hectares and accounts for 40% of the world's agricultural production. This sector provides food and income for around 1 billion people worldwide (FAO, 2015).

Africa holds a crucial share of the world's food security, with nearly 600 million hectares of uncultivated land, or 60% of the global total (MERLET, 2013). Since independence, the land issue has become crucial in many African countries. It lies at the intersection of multiple issues, and numerous studies have shown that the right to land can be an instrument of development. The importance and difficulties associated with land ownership in the countries of the South are no longer in dispute (MALALA, 2014).

Madagascar, the world's fourth largest island, devotes a large part of its territory to agriculture, thanks to its abundant water resources. More than half the population lives in rural areas (PERRINE, 2022). Agriculture and livestock farming play a key role in both human and economic terms. The country has around 36 million hectares of arable land, of which only 3 million, or less than 10%, are farmed. It also has 1.5 million irrigable hectares, of which 1.1 million are equipped (FAOSTAT, 2009).

Land control is one of the main development challenges in Madagascar. The regulation of land ownership is an essential link in the development chain, particularly in a predominantly agricultural country where the population is strongly linked to land management (EDDY, 2018). At Madagascar, owning land is considered a determining factor of wealth, which confers great importance to land for the Malagasy (MINOHERILALA, 2022).

It is in this context that our topic is entitled "**administrative procedures and conditions of access to land for agricultural and livestock activities: the case of the urban commune of Diego Suarez**". The analysis of this topic aims to gain a better understanding of the conditions to be met and the procedure to be followed for access to state-owned land. This work seeks to answer the following questions:

- How do you get access to land in Madagascar?
- By what means and under what conditions can the population of the Urban Commune of Diégo-Suarez gain access to State-owned land?

The general objective is to regulate private and public land use to ensure a balance between private, public and environmental interests. Specific objectives include:

- Exploring administrative procedures for obtaining title to private and public land;
- Identify suitable methods and conditions for accessing state-owned land for agro-ecological activities.

The research questions are as follows: What documents are required to apply for access to state-owned land? What are the conditions required to obtain state land? Where should the application be submitted?

The hypothesis formulated is that access to state-owned land in Diégo-Suarez is hampered by complicated administrative procedures and land conflicts, hampering economic development. Procedural reform and an improved legal framework could facilitate access to land and foster sustainable, inclusive development.

In order to answer these questions, we'll look at some general points in the first section. Then, in the second part, we will describe the materials and methods used. In the third part, we'll look at the results obtained and discuss them. And finally, we'll close with a conclusion.

2. GENERAL

2.1 LAND STATUS AND ACCESS IN MADAGASCAR

2.1.1 Land status

2.1.1.1Legal framework

Framework law N°2005-019 of October 17, 2005 establishes the principles governing land status in Madagascar (**Source**: www.observation.foncier.com). Under the terms of this law, land in the Republic of Madagascar is divided as follows:

2.1.1.2State-owned land

These lands are divided into two categories: the public domain and the private domain of the State.

- **Domaine Publics (DP)**

The PD includes all natural or man-made assets for which the State or a decentralized authority is responsible for protecting and managing in the collective interest. The PD is inalienable, unseizable and imprescriptible. It includes elements such as roads, lakes, rivers, beaches, seaports and railroads. This domain is governed **by law 2008-013 of July 23, 2008 and its implementing decree N° 2008-1141 of December 1, 2008**(MINOHERILALA, 2018).

- **State-owned private domain (DPE)**

The ECD, whether real estate or movable property, covers all goods and rights likely to be privately owned by virtue of their nature or purpose. It is governed by **law N° 2008-014 of July 23, 2008 and implementing decree N° 2010-233 of April 20, 2010**. The Domaine Privé de l'Etat is divided into two parts, such as the titled Domaine Privé de l'Etat and the untitled Domaine Privé de l'Etat without development (Source : Inspecteur des domaines : RABEMIARISOA Olivia Dany, 2024).

- ✓ **State-owned private domain Titled**

There are two types of titled State Domain: the titled State Private Domain affected and the titled State Private Domain unaffected. Restricted State Private Domain is untouchable, while unrestricted titled State Private Domain can be acquired by purchase, free or onerous concession, simple lease, or made available for harvesting or exploitation rights by any natural or legal person.

- ✓ **Untitled State private domain without development**

This type of land is not developed in any way, and a request for acquisition can be made to the land department.

2.1.1.3Private land

Private land comprises titled and untitled private property.

- ❖ **Private titled property (PPT) or cadastral property**

Private Titled Properties are governed by Ordinance No. 60-146 of October 3, 1960 on land registration and Decree No. 60529 of October 3, 1960, regulating the application of this ordinance.

- ❖ **Propriété Privée Non Titrée avec Mise en Valeur or PPNT**

Untitled private property is governed by **law N° 2006-031 of November 24, 2006, setting out the legal regime for untitled private land ownership, and by decree N° 2007-1109 implementing this law**. This type of ownership applies to all urban and rural land that is occupied but not registered in the land register, that does not form part of the public or private domain of the State or a decentralized authority, and that is not located in an area subject to a special status.

2.1.1.4Land with special status

These lands are subject to a specific legal protection regime, such as nature reserves, special economic zones and agricultural investment zones.

The Commune Urbaine de Diégo Suarez is a decentralized territorial authority. These "Collectivités Territoriales Décentralisées" are legal entities under public law with financial and administrative autonomy. They are freely administered by municipal councils in accordance with the terms and conditions laid down by the law and regulations in force. Subsequently, it brings together all the State's decentralized services at regional level.

As long as it is a "Collectivité Décentralisée", the commune depends structurally on the Ministry of the Interior and hierarchically on the Ministry of Decentralization. It comprises districts whose territorial limits coincide with those of the former sub-prefecture (in the case of the territorial collectivity). It is a public authority with an essentially economic and social vocation. It directs, drives, coordinates and harmonizes the economic and social development of its entire territory. As such, it is responsible for planning, committing the territory and implementing all development actions (TANDRA, 2016).

2.1.2 General information on access to land and organization chart of cirdoma Diego-I personnel

2.1.2.1General information on access to land

Access to land, although theoretically guaranteed by a legal framework, remains difficult for many people, due to suspended or ineffective measures. Organized agricultural credit, although promising, is rarely accessible due to insolvency, forcing people into excessive debt. Day-to-day constraints such as lack of inputs, affordable credit, and poor land distribution exacerbate this precariousness, threatening people with the loss of their own land. Moreover, the complex interplay between customary law and modern land law adds further uncertainty to the situation (RAMAROLANTO, 1989).

2.1.2.2 Diégo-I cirdoma staff organization chart

The cirdoma Diégo-I staff includes: two (2) estate inspectors, one (1) accountant, eight (8) editors, one (1) archivist and one (1) janitor.

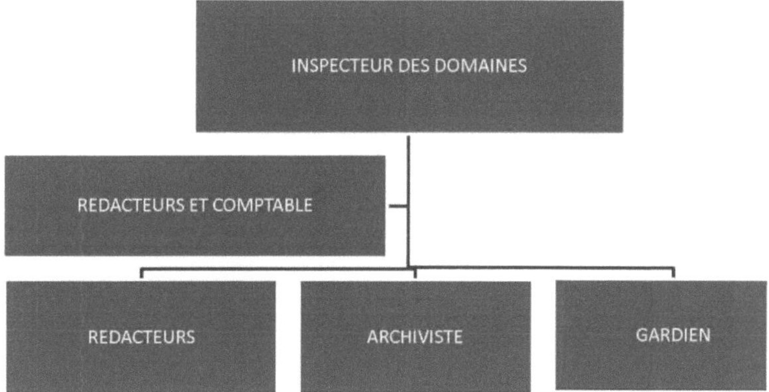

Figure 1: Organization chart of Diégo-I cirdoma staff, 2024

2.2. GEOGRAPHICAL AND ECOLOGICAL CHARACTERISTICS OF THE STUDY AREA

2.2.1. Location of study area

The urban commune of Diégo Suarez, the regional capital of DIANA, is an economic and administrative command town in the far north of Madagascar. It brings together various sectors of regional activity, including tourism, and is an important host town in the region. According to GPS coordinates, the heart of the town lies between 12° 16'984 south latitude and 49° 17'384 east longitude.

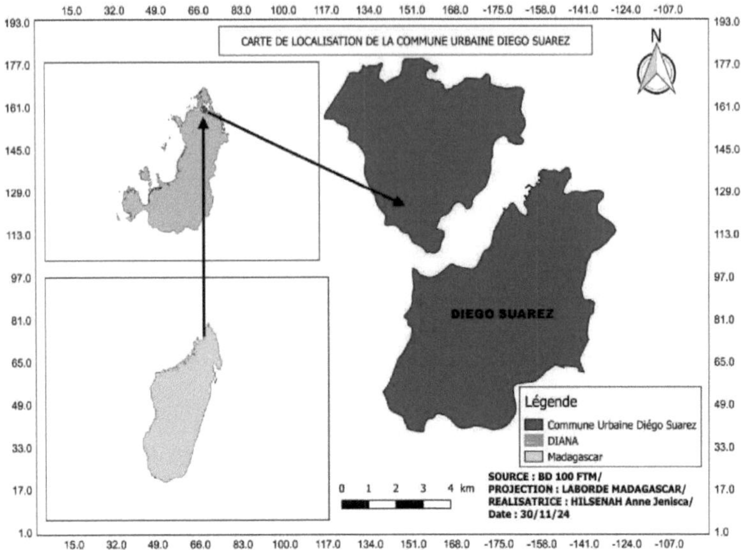

Figure 2: Study area location map

2.2.2. Natural environment

2.2.2.1. Physical environment

The town of Antsiranana is located at Cap d'Ambre (Tendron'i Bobaomby) in northern Madagascar. It is bordered to the west by the Mozambique Channel and to the east by the Indian Ocean. Its surface area extends over 47km^2 and comprises 25 districts.

❖ Climate

Antsiranana's climate is tropical, characterized by an alternation between :

- A long, cool, dry season (April to November), marked by the varatraza winds;
- A hot, rainy, shorter season (December to March), influenced by the

monsoon.

Table 1: Temperature variation by month, 2021

Month	Min (°C)	Max (°C)	Average (°C)
January	23	28,6	25,2
February	23	28,5	25,1
March	22,2	28,8	25,4
April	23	29	24,8
But	22,2	28,6	23,5
June	20,8	27,4	22,7
July	20	26,9	23,3
August	19,8	27	22,7
September	20,1	27,9	23,3
October	21	29,1	24,4
November	22,1	29,7	25,3
December	22,9	29,6	25,6

Source : www.fr.Wikipedia.org/wiki/Antsiranana

❖ **Precipitation**

Annual rainfall is around 1,195 millimetres, with a maximum of 340 millimetres in January, the wettest month (Source: climat-Antsiranana (MADAGASCAR)).

❖ **Soil**

The diversity of soils in the region is linked to variations in climate and relief. The main distinctions are :

- ❖ Advanced (lateritic) soils, rich in free alumina, are found in the middle part of the region;
- ◆ Red soils, derived from basalt and devoid of organic horizons, are

characterized by a dark red or yellow-red hue (SEGALEN, 1956).

- **Relief**

Relief is the set of irregularities in the earth's surface. The city is surrounded by hills and mountains, notably the amber massifs to the west, which are a nature reserve. The relief is also marked by coastal plains and mangrove swamps, and encompasses all the irregularities of the earth's surface: mountains, hills, plateaus, plains, etc. The state of the relief depends on various factors. The state of relief depends on various factors, including the nature of the rocks, the environment, climate, the type of erosion and its duration.

The far north is made up of a volcanic relief of basaltic cast type and a sedimentary relief. As far as the sedimentary relief is concerned, the limestone rocks have given rise to landforms fairly comparable to those of the basaltic rocks, as the limestone zone is entirely under a dry climate, with a characteristic tabular relief bordered by escarpments. The sandstone hills, on the other hand, are deeply eroded in all its forms, as evidenced by the numerous rip-outs. The city is surrounded by hills and mountains, notably the Ankarana massif, and the coastline is marked by beautiful bays and beaches. This geographical diversity influences the local economy, particularly fishing and tourism (BOTRA, 2022).

- **Hydrography**

Most rivers have their source in the Ambre massif, culminating at 1400 m. They are characterized by torrential flow at altitude and become slower as they approach the sea. The oval-shaped alluvial plains are often home to mangroves. In the Bobaomb peninsula, there is little to say about the rivers, which are very short and practically dry in the dry season. Their course is generally east-west (SEGALEN, 1956).

2.1.2.3Biological environment

- **Vegetation**

The vegetation of Diégo Suarez is influenced by its tropical climate. It includes :

- Tropical forests;
- Mangroves along the coast ;
- Savannahs and semi-arid zones. The surrounding parks, such as Ankarana National Park, are home to a unique biodiversity, with many endemic plant species (Source: flora in northern Madagascar).

- **Fauna**

The fauna of Diégo-Suarez includes a wide variety of endemic species at the many tourist sites. These include lemurs, mammals such as bats, and birds such as pigeons and raptors. Marine fauna, including tropical fish and sea turtles.

2.2.3. Population and main activities

2.2.3.1. Population

In 2005, the Commune Urbaine d'Antsiranana had a population of around 135228, with an annual demographic growth rate of 3%. Migratory flows account for 20% of the total population, and experts for around 2%. The average density is 30 inhabitants per hectare (AMBINIAINA, 2016).

2.2.3.2. Main activities

The region's main activities include :

- Agriculture and livestock ;
- Fishing and aquaculture ;
- Trade and transport.

- **History**

Diégo-Suarez was a village of a dozen fishermen. In the 15th century, Diégo-Suarez was discovered by two Portuguese navigators, Diégo Diaz, who discovered the big island on August 10, 1500, and Fernando SUAREZ, who was the first to discover the lease in 1534. In February 1506, Admiral Herman Suarez recognized the place, and the town acquired the captain's first name and the admiral's name: DIEGO-SUAREZ.

From 1885, the town of Diégo-Suarez came under French control and grew in importance thanks to the French Navy, which set up an arsenal and several barracks. When the military agreements expired, the port was transformed into a shipbuilding and repair yard, one of the largest in the Indian Ocean.

The Diégo-Suarez town hall was built and installed in 1954-1962 during the colonial era, when the late SAUTRON was the town's first mayor.

From 1960 to 1976, the town was a Commune Urbaine, and at the same time the capital of the province of Antsiranana. But when the decentralized authorities were set up, it became a Fivondronampokotany, and the town was called Antsiranana I. So from 1995, the town became Commune Urbaine de Diégo-Suarez again, in application of decree n°95381 of May 26, 1995 (TANDRA, 2016).

3. MATERIALS AND METHODS

3.1 MATERIALS

The materials used for this study include :

- A notebook for taking notes ;
- Pens to record information;
- Survey sheets to gather information on administrative procedures and land access conditions
- GIS software for map production ;
- Internet access to books and theses
- A computer for data processing

3.2 METHODS

The method adopted for this study is based on a scientific approach integrating bibliographical research and field surveys at the Direction de l'Aménagement du Territoire and the urban commune of Diégo-Suarez.

3.2.1 Bibliography search

This stage consists of researching, identifying and analyzing relevant documents related to the study topic, by developing a rigorous search strategy. We consulted resources available in the university library as well as online databases. The library of the University of Antsiranana (UNA) proved to be a valuable source for an initial, more traditional methodological approach.

3.2.2 Survey

Semi-structured interviews were conducted with officials from the Commune Urbaine de Diégo-Suarez and the Direction de l'Aménagement du Territoire (see Appendix 01). This qualitative method provides in-depth information on the practices and perceptions of the players involved **(DUPONT, (2022)).**

3.2.3 Data collection

A structured questionnaire was developed to gather information on the conditions and procedures for accessing public and private land for agricultural and livestock activities in the urban commune of Diégo-Suarez. Data collection is a crucial stage in the research process, guaranteeing the reliability and validity of the results (**MARTIN, (2021)).**

3.2.4 Data processing

The data collected was entered, organized and analyzed using data processing and mapping software, producing usable results relevant to the study. This stage was carried out in compliance with current methodological standards **(LEROY, (2020)).**

4. RESULTS AND DISCUSSION

4.1 RESULTS

4.1.1 Tenure conditions for DPEs and DPs in Diégo-Suarez

The conditions for granting Established Property Rights (EPR) and Property Rights (PR) in Diégo-Suarez are designed to ensure that only applicants meeting specific criteria are granted access to land. These criteria are designed to protect land rights and ensure responsible management of land resources.

Firstly, Malagasy nationality is a fundamental prerequisite, underlining the importance of land ownership for the country's citizens. The minimum age requirement of 18 ensures that applicants are legally capable of entering into contracts and making decisions concerning property. In addition, restrictions on persons who have been deported, placed under house arrest or sentenced to terms of imprisonment of one year or more are designed to prevent abuse and ensure that applicants have no criminal record that could compromise their ability to manage land responsibly. Article 11, which imposes a ten-year waiting period, reflects a desire to stabilize the land situation and encourage sustainable land use. This waiting period also enables the authorities to verify applicants' compliance with legal requirements and ensure that they have a legitimate interest in the land applied for. It can also serve to dissuade land speculators who might seek to acquire land with no intention of developing productive activities on it.

Before submitting an application, applicants must assess the situation of the land concerned. This assessment is crucial, as it enables applicants to understand the characteristics of the land, its legal status, and any environmental or regulatory constraints. A good knowledge of the land situation can also help to avoid future disputes and ensure that the land will be used in accordance with current legislation.

Finally, submitting an application to the estates department is an essential administrative step. This process involves the preparation of supporting documents and the submission of a complete dossier, which may include development plans, environmental impact studies and other relevant information. The estates department plays a key role in assessing applications and issuing title deeds, ensuring that land rights are allocated fairly and transparently.

4.1.1.1Land ownership procedure in the Urban Commune of Diego-Suarez

- **Accession of an ECD plot**

Under Malagasy land law, management of the State's private domain is governed by the specific nature of the land, whether titled or not. Land transfer procedures enable certain categories of individuals to acquire plots of land, thereby recognizing the rights of occupants who have personally developed the land in question.

The stages of acquisition of the State's private domain in the urban commune of Diégo-Suarez include :

Table 2: Acquisition stage of the CUDS state-owned private domain

PREPARATION OF ADMINISTRATIVE STAGES PLANS	
- Location plan - Regular plan for unregistered land - Official plan for unregistered land or registered	- Prior contact with the topographic - Submission of application accompanied by consignment (200 Ar/Ha with a minimum of 5000Ar) - Inventory of fixtures (CEL) - Advice from the relevant technical departments - Registration or division in the name of the Malagasy State - Collection of state provision - Draft deed and send for approval - Registration of the deed in the land register

- **Accession of a DP lot**

In principle, public domain land is inalienable, unseizable and imprescriptible. However, certain parts of the DP may be allocated for private use, by means of a concession contract or temporary occupation authorization. A special authorization may also be issued in favor of the applicant, for a renewable period not exceeding 30 years.

The procedure for accessing DP land follows the same steps as those for accessing state-owned land, but with some specific features:

- At the reconnaissance stage, the presentation of public works is compulsory. This ensures that the planned projects are in line with the community's infrastructure needs.

- An occupancy decree must be drawn up and approved by the Minister of Lands, which formalizes the authorization to occupy the land.
- Instead of signing the deed and the attached plan, the applicant must produce a submission undertaking to comply with the provisions of the by-law. This creates a legal and binding framework for the applicant.

For areas between 250 Ha and 2,500 Ha, acquisition requests must be submitted for prior approval by the Minister responsible for land. This allows for a more rigorous assessment of acquisition projects, ensuring that they meet the needs of sustainable development and land use planning.

For areas larger than 2,500 hectares, a specific procedure is put in place. This procedure may include environmental impact studies, public consultations, and detailed assessments of the socio-economic implications of the acquisition. The steps to be followed for these large areas may include :

1. **Preparing a complete file**: The applicant must provide a detailed file including technical studies, development plans and economic justifications.
2. **Stakeholder consultation**: Meetings can be organized with local communities and authorities to discuss potential impacts and gather opinions.
3. **Expert assessment**: The file will be examined by land-use planning and environmental experts to ensure that the project complies with current standards.
4. **Final approval**: After evaluation, the dossier will be submitted for final approval to the Minister in charge of land tenure, who will make the decision taking into account the recommendations of the experts and the opinions of the stakeholders (table 3).

Table 3: Stages in the process of acquiring a large plot of land

ETAPES	PIECES A FOURNIR	RESULTATS ATTENDUS	RESPONSABLE
ANALYSE DU PROJET	Dossier de procédure avec son business plan	-Impact socio-économiques du projet (au niveau local, régional et national etc.) -Délai (cycle de culture, …) -PV de réunion de comité -Décision de comité : positive ou non -S'il le faut soumission du dossier au conseil des Ministres.	-Comité interministériel institué pour chaque projet, par décision du Ministère chargé du foncier. -Représentants des CTD concernées.
Notification du requérant		Poursuite de procédure ou abandon du projet.	Le comité interministériel.
Dépôt de la demande du dossier du terrain	Busines plan, avis favorable du ou des ministères sectoriels concernés, PV du comité interministériels note d'approbation du comité etc.	Récépissé de la demande	-Demandeur -Ministère Chargé du Foncier
Délivrance autorisation de prospection (en deux langues)		Autorisation dont la publicité est à la charge du demandeur	Ministère Chargé du Foncier
Enquêtes administratives		Terrain disponible et quitte de toutes charges	Demandeur, Région, CTD, tous services techniques concernés. (au frais du demandeur)
Remise des résultats de la prospection, des avis des autorités régionales et dossier de demande d'acquisition au VPDAT	Pièces de procédure prévues par la loi n°2008-014 et autres jugés utiles (NIF, RCS, Statuts, etc...)	Accord de principe	-Demandeur -Ministère Chargé du Foncier
Instruction de la demande auprès des services déconcentrés de la situation de l'immeuble	Procédure normale prévue par la loi n°2008-014	Décision de principe acquise, immatriculation de l'immeuble, provision domaniale payée, acte et cahier des charges rédigés	Cirdoma, cirtopo, tous les services techniques concernés
-Approbation du projet d'acte -Enregistrement de l'acte	-Dossier complet -Projet d'acte dument approuvé	-Acte approuvé -Acte enregistré	-Ministère Chargé du Foncier -Centre fiscal, Demandeur
Ampliation conforme de l'acte, inscription de droit au bail	Dossier complet	Droit au bail publié	-Circonscription Domaniale et foncière -Demandeur

Some acts require consultation with a certain number of bodies, some requests or decisions must be made on the basis of an imposed model, and in some areas inquiries are required to ensure that the authorities are better informed (town planning, environmental permits). The authority must take into account any complaints made during these inquiries. Certain legal provisions require opinions, or even assent. In certain cases, a report assessing the impact of the project must be drawn up for draft laws and draft decrees to be deliberated by the Council of Ministers on the potential effects of any draft regulation...on the economy, the environment, social aspects and administrations.

❖ **Required documents**

The procedure for acquiring land in Diégo-Suarez requires the submission of a set of specific documents, each of which plays a crucial role in the evaluation and validation of the application. These documents aim to guarantee the transparency, legality and conformity of acquisition requests, while protecting the rights of the parties involved.

-Plan with compulsory preliminary survey: This document is essential for pinpointing the precise location of the land in question. It enables the authorities to assess the geographical location and characteristics of the land, thus facilitating the review process.

-Official plan for registered or cadastred land: For land that has already been registered, this official plan is needed to confirm the legal status of the land and ensure that there are no outstanding ownership disputes.

-Regular plan drawn up in accordance with topographical service standards: This plan, drawn up by a sworn surveyor, must be accompanied by a procès-verbal (PV) describing the boundaries. This ensures that the dimensions and boundaries of the plot are clearly defined, minimizing the risk of future conflicts with other owners.

-Attestation de non mise en valeur issued by the Fokontany chief: This document certifies that the land is not currently being farmed or developed, which is crucial in determining the possibility of acquisition. It also ensures that the land is available for new projects.

-CSJ (Certificat de Situation Juridique): If the land is registered and cadastralized, this certificate is required to confirm the legal status of the land and ensure that there are no encumbrances or restrictions that could affect the acquisition.

- Application on forms provided by the administration: Submitting an application in duplicate, duly completed, is a standard administrative requirement that keeps an official record of the application. -Certified photocopy of National Identity Card (CIN): This document is needed to verify the applicant's identity and ensure that he or she meets the legal requirements to apply for acquisition.

-Duly legalized power of attorney: If the application is made by a person acting on behalf of another, a legal power of attorney is required to guarantee that the representative has the necessary authorization to act on behalf of the applicant.

-Guardianship order or letter of undertaking from the legal guardian for minors: For minors, it is imperative to include legal documents attesting that the guardian has the right to manage land affairs on behalf of the child.

-Documents for companies: Companies must provide their articles of association, an extract from the commercial register and a mandate from the legal representative. This verifies the company's legitimacy and ensures that the person submitting the application has the necessary authority to do so.

-Deliberation of the deliberative body for applications submitted by CTDs (Collectivités Territoriales Décentralisées): This document is required to prove that the application has been approved by the competent body of the local authority, thus guaranteeing that the acquisition complies with local decisions.

-Investment project for land over 10 Ha: For applications concerning large plots of land, a detailed investment project is required. This project must demonstrate how the land will be used and the benefits it will bring to the community, which is essential to justify the acquisition of large parcels of land.

-In short, thoroughness in the preparation and submission of these documents is essential to ensure successful land acquisition. Each document plays a key role in the appraisal process, ensuring that the rights of applicants are respected while protecting the interests of the State and the community. This systematic approach contributes to more transparent and accountable land management, fostering sustainable development in Diégo- Suarez.

- **Role of the municipality**

Assist with demarcation, inventory and demarcation operations.

4.1.1.2 Guide to Access and Regulation of Mining Activities

- **Presence of natural resources**

A- Access to land use: mining perimeter

Ownership of the land is linked to possession of a mining permit, in accordance with Law 99-022 of August 19, 1999, amended by Law 2005-021 of October 17, 2005.

Mining perimeter: is defined as the square or set of several contiguous squares joined at the sides which are the subject of a mining permit or application for a mining permit.

The mining permit holder is obliged to inform the landowner of his right to occupy the portion of the property covered by the mining permit, and to enter into a lease agreement specifying their respective rights and obligations. In the absence of a lease agreement, the permit holder may be entitled to compensation , and in the event of disagreement, the matter may be referred to the civil courts.

A. Procedure and documents required to apply for a mining permit

- ✓ **Procedure**

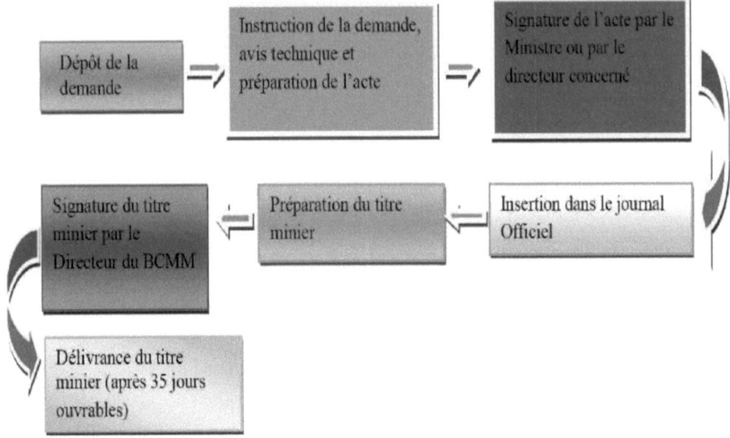

✓ Documents to be supplied

Documents required to apply for a mining permit include:

- Three (03) passport photos
- Application form duly completed and signed
- AERP if applicable
- Legalized standard plan
- Commitment letter for PEE and EIE
- Locator map and map
- Power of attorney
- Bulletin N°3 (less than 3 months old)
- Certified copy of CIN and Certificate of Residence (less than 3 months old)
- Certified copy of current year's professional mining card
- Certified copy of commercial register certificate
- Certified copy of tax return
- Certified copy of company statutes

B- Housing

✓ **Verification of land status and declaration of work**

He must ensure that the land is suitable for building and that it complies with local regulations (PLU - Plan Local d'Urbanisme), and check for easements, rights of way and other restrictions. For minor works, a prior declaration may suffice.

- ✓ **Obtaining a PC and necessary documents**

-Obtain

To build legally in Diégo Suarez, you need a building permit, which is the result of a preliminary project. This permit is the result of a preliminary project and requires several stages that can take several weeks, even months, although the town hall undertakes to validate files within a month. Once the permit has been obtained, the work must be completed within one year.

-Required documents

Documents required for the building permit application include: Alignment

- 5 topographical plans
- 3 plans of the house to be built
- 3 layout plans
- 1 legal certificate less than 3 months old
- 1 handwritten request to the town hall and the Fokontany chief

All these documents must be enclosed in a cardboard folder. Depending on the Fokontany, the receiver should normally provide a numbered receipt. The last two digits of the number indicate the year of filing.

There are 2 types of PC:

- **Information certificate**: gives the planning regulations for a given plot of land
- **Professional certificate**: provides information on the feasibility of a project

The deadlines are 2 months for building permits for single-family homes and 3 months for other types of projects.

- ✓ **Compliance with building standards, network connection, insurance**

You'll need to apply for the necessary permits (water, electricity, etc.) and ensure that the construction complies with safety, accessibility and environmental standards. The owner must pay the property tax, inform the town hall of the occupation of the land and, if necessary, register the residence, while remaining attentive to property regulations or specific neighborhood rules.

4.1.2. Conditions of Access to State Land for Agroecological Activities

❖ Current land use situation

In the CUDS, agriculture and livestock play a central role in land use. In terms of contribution to the Diégo Suarez economy, livestock farming dominates. Although present, livestock farming has not reached the same level of development as agriculture in the CUDS.

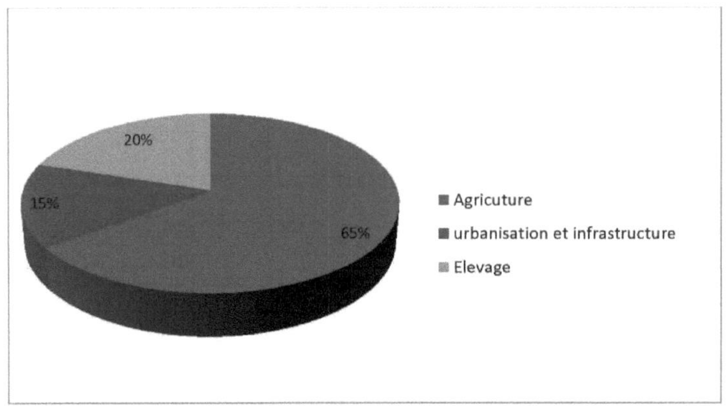

Figure 3: Percentage of land use in Diégo-Suarez

In the CUDS, urbanization occupies a significant proportion of the land. Urban, industrial and commercial zones, particularly around the port and airport, are taking up more and more of the available land, around 15%. Diégo-Suarez is favorable to agriculture, with sugarcane, rice, vegetables, fruit and other crops accounting for around 65%. Livestock farming remains a more modest sector compared to agriculture. Cattle rearing (mainly zebus and goats) is also present, but generally occupies a smaller share of land than agriculture, around 20%.

❖ Administrative steps to follow and documents to provide for livestock and agriculture

The use of land in Diégo-Suarez requires compliance with certain administrative steps and regulations, as shown in the following table:

Table 4: Administrative steps for an agro-ecological activity

PROCEDURES A SUIVRE	PIECES A FOURNIR	LIVRABLES
Demande d'autorisation d'installation auprès de la Direction Régionale en charge de l'Elevage et de l'agriculture	- Autorisation du Fokontany et/ou de la commune - Accord des riverains - Renseignement concernant l'éleveur, activité envisagée, emplacement et ma taille de l'exploitation	Autorisation d'installation,
Demande de numéro d'identification de l'exploitation et carte d'éleveur ou carte de production auprès de la Direction Régionale en charge de l'Elevage et de l'agriculture	- Renseignement concernant l'éleveur et ses activités - Autorisation d'installation	Carte éleveur Carte de production
Demande d'autorisation d'implantation auprès de la commune	- Statut juridique - Plan d'implantation - Situation juridique ou contrat de location - Fiche de projet spécifiant les activités	Autorisation d'implantation délivrée par la commune
Demande de permis environnementale auprès de l'ONE (Office Nationale de l'Environnement)	- Fiche tri remplie - Autorisation d'implantation délivrée par commune - Documents du projet avec facture préforma de la liste des investissements des matériel et équipement - Plan d'aménagement approuvé par le service de l'environnement	Permis environnementale délivrée par ONE avec cahier de charge ou Autorisation environnementale délivrée par la Ministre en charge de l'Elevage et de l'agriculture avec cahier de charge ou certificat de conformité délivrée par ONE avec cahier de charge
Demande d'autorisation d'installation adressée auprès de la Direction technique	- Autorisation d'implantation - Statut juridique - Plan d'aménagement - Situation juridique - Documents descriptif du projet - NIF STAT - Permis environnementale - Avis favorable de l'entité en charge de l'amélioration génétique en cas d'espèces ou race nouvellement introduite - Quittance des paiements des droits au compte du fonds de l'élevage	Plan d'aménagement approuvé par le service environnemental Autorisation d'installation standard délivrée par la Direction technique
Installation proprement dites		Rapport de visite
Demande d'autorisation d'exploitation adressée à la Direction technique	- Autorisation d'installation - Document de projet avec plan d'aménagement et compte d'exploitation prévisionnel y compris listes des investissements en matériel et équipement	Autorisation d'exploitation standard délivrée par la Direction technique
Exploitation		Rapport de suivi
Demande d'agrément	- Autorisation d'exploitation - Documents de l'exploitation : -Production -Sanitaire : programme de surveillance zoo sanitaire, mouvement des animaux, mise en place de la biosécurité -Agrément sanitaire délivré par des services vétérinaires	Agrément délivré par la direction technique à la demande de l'exploitant

To set up a production unit, an application must be approved by the Fokontany/Commune and local residents. The file is then submitted to the DRE to obtain an identification number, production card and breeder. After a positive assessment of the EIA or PREE, the ONE or MEA issues the PE or AE with specifications. For existing units, a certificate of conformity replaces the AE. In the event of diversification or extension, a new permit application is required, and an environmental discharge is required if operations are discontinued. Technicians carry out a site visit to check compliance before approving the plan, then a payment of installation fees is made to the Fond de l'Elevage. A second inspection visit is required before the operating permit is issued.

4.2. DISCUSSION

4.2.1. Complex land access procedures

Access to land in Madagascar, and particularly in the Commune Urbaine de Diégo-Suarez, is marked by an administrative complexity that can discourage many applicants. Acquisition procedures vary considerably depending on whether the land is state-owned or private, and each type of land imposes specific requirements. Malagasy land laws, although intended to regulate access to land, are often perceived as a bureaucratic labyrinth.

The administrative procedures required, such as applying for authorization to set up, obtaining a farm identification number, and applying for an environmental permit, are often long and tedious. ANDRIAMANANTENA (2O18) points out that the lack of transparency and coordination between the various public services further complicates access to land. Citizens often find themselves faced with long waiting times and seemingly unattainable administrative requirements, which can create a sense of frustration and injustice.

In addition, the slowness of decision-making processes, as noted by RANDRIAMAMONJY (2020), acts as a brake on local economic development. Potential entrepreneurs and farmers are thus discouraged from committing to investment projects, which is detrimental to the region's economic growth. The complexity of land access procedures, coupled with a lack of clarity in regulations, represents a major obstacle to individual initiative and innovation in the agricultural and agro-ecological sector.

4.2.2. Unequal access and development issues

Inequalities in access to land in Madagascar, particularly in Diégo-Suarez, raise important concerns about social justice and sustainable development. RAVONINJATO (2019) highlights the impact of land laws on the rural population, where increasing urbanization competes with agricultural needs. Although current legislation aims to protect citizens' rights, inequalities persist, exacerbated by complex administrative procedures and requirements that do not always take account of local realities.

Vulnerable populations, particularly small-scale farmers and young entrepreneurs, are often the hardest hit by these inequalities. Access to land, a key factor in economic development and food security, remains limited for those who lack the resources or knowledge to navigate the administrative system. This creates a vicious circle where the most disadvantaged are excluded from development opportunities, reinforcing economic and social disparities.

To remedy this situation, reforms are needed to improve the transparency and efficiency of administrative procedures. This could include simplifying acquisition procedures, setting up support mechanisms for vulnerable populations, and promoting better coordination between different public services. By promoting equitable access to land, Madagascar could not only boost its economic development, but also strengthen social cohesion and community resilience in the face of environmental and economic challenges.

5. CONCLUSION AND RECOMMENDATIONS

Access to land in the Commune Urbaine de Diégo-Suarez, particularly for agricultural and livestock activities, is a complex process. It involves numerous administrative procedures and legal requirements aimed at regulating land use while ensuring compliance with environmental, social and economic standards. These procedures, although adapted to the local context, often follow guidelines common to many countries.

Farmers must comply with a series of regulations, whether they are working on land in the State's private domain or on other types of landholding. Administrative procedures require the presentation of several supporting documents, such as mining permits, building authorizations or environmental impact studies. Access to such land also depends on the eligibility of operators, who must prove their qualifications and their ability to respect the rules of sustainable natural resource management. To improve land management in Diégo-Suarez :

- **Simplify administrative procedures**: Reduce delays and formalities, especially for large-scale operations.
- **Enhance transparency**: Create a centralized platform for land applications and mining permits.
- **Encourage sustainability**: Require rigorous environmental assessments for agricultural and mining projects.
- **Train local players**: Raise awareness of land and environmental issues among applicants and local authorities.
- **Promoting sustainable agriculture**: adapting administrative procedures to agro-ecological projects.

These measures are designed to make access to land more equitable, ensure compliance with environmental standards and promote harmonious, sustainable development.

6. REFERENCES BIBLIOGRAPHIES

-AMBINIAINA, H.J., 2016. PROMOTION DE L'ASSAINISSEMENT ET DE LA GESTION DE L'ENVIRONNEMENT URBAIN "CASE DANS LA COMMUNE URBAINE ANTSIRANANA", page: 47. Dissertation for the Master of Engineering Water Sciences and Techniques (M.I.S.T.E).

-ANDRIAMANANTENA (2018), LE FONCIERA MADAGASCAR, Page: 511, 56.

-BOTRA, S., 2023. AGRICULTURE DURABLE ET LA SECURITE ALIMENTAIRE DANS LA COMMUNE URBAINE DE DIEGO-SUAREZ ; Université d'Antsiranana, page : 30.

-DUPONT, J. (2022). *Méthodes d'enquête qualitative en sciences sociales.* Paris: Éditions de la Recherche.

-EDDY, R., 2018. ACQUISITION DE TERRAINS DU DOMAINE DE L'ETAT PAR LES ETRANERS, page: 37. Dissertation in view of obtaining the Master's degree in law.

-FAO (Food and Agriculture Organization of the United Nations, 2022. Land Tenure and Rural Development, FAO Land Tenure Studies 3.

-FAO and TECA, 2015. Training manual for organic agriculture.

-HERINIAINA, R., 2022. Marché foncier et accès à la terre des migrant : une analyse institutionnelle dans une commune rurale à Madagascar, page : 8.

LEROY, S. (2020). *Data analysis: Methods and tools.* Marseille: Éditions Scientifiques.

-MALALA, (2014). La pertinence du nouveau système de droit foncier de Madagascar (la réforme foncière de 2005) thesis; Paris 1.

-MANUEL PROCEDURES EN ELEVAGE, 2016, page : 4-14.

-MARTIN, L. (2021). *Techniques de collecte de données en recherche appliquée.* Lyon: Presses Universitaires de Lyon.

-MERLET, 2018. C2A notes Agriculture and food in question, page : 2.

-MINOHERILALA, L.F.S., 2018. BORNAGE DE MORCELLEMENT D'UN TERRAIN SIS A AMBODISAHA DANS LA COMMUNE URBAINE AMBOHIDRATRIMO, page: 8-9. Final dissertation for the licentiate degree in science and technology in geographic and land information.

-NELLY, R., RALAMBONDRAINY. First Honorary President of the Supreme Court of Madagascar. ACCES A LA PROPRIETE FONCIERE A MADAGASCAR, page : 20,43.

-PERRINE, B., 2022. Marché fonciers et accès à la terre des migrants: dans l'Ouest de Madagascar: opportunités et contraintes. Economie Rurale (381): 79 -93.

-RALANTORATIARAY, 1989. ACCES A LA TERRE EN DROIT RURAL MALGACHE, maitre de conférences à Madagascar, Directeur du center d'Etudes Rurales, page: 38(3).

-RAMARISON, T.N., 2022. ETUDES HYDROGEOLOGIQUES DE LA REGION DIANA, Mémoire de fin d'étude en vue d'obtention du diplôme de licence en ingénierie pétrolière, page: 21.

-RANDRIAMAMONJY, 2020. Le système Foncier Malgache: défis et perspectives, page 112.

-RAVONINJATOVO, 2019. Land tenure and agriculture in Madagascar: problems and solutions, page 78.

-SEGALEN, P., 1956. Pédologie de Diégo-Suarez, page: 269-270.

-TANDRA, N.E., 2016. L'EXECUTION DE LA PROCEDURE DE PAIE DU PERSONNEL AU SEIN DE LA COMMUNE URBAINE DE DIEGO-SUAREZ, page: 3. Graduation thesis for the Higher Technician Diploma in Business Administration.

7. WEBOGRAPHY

http://www.Sandrinal3fle.Wordpress.com/2017/12/02/nord/. Accessed on 09 october 2024

http://www.fr.Wikipedia.orgg/Wiki/Antsiranana. Retrieved October 01, 2024

http://www.fr.climate-data.org/afrique/Madagascar/diégo-suarez/diego-suarez-3099/Consulted October 06, 2024

https://www.climatsetvoyages.com. Retrieved October 07, 2024

www.observation.foncier.com. Retrieved November 14, 2024

www.Droit.Afrique.com. Retrieved November 18, 2024

8. APPENDIX 01: Survey form

Type of survey individual interview

What are the administrative procedures for titling state-owned land?

What documents do I need to apply for access to a plot of land?

Where can I apply for access to a plot of land?

What is the status of land in Diégo-Suarez?

Who makes the decision to authorize use of the land?

Who signs the contract?

What are the conditions for access to the land?

What are the requirements for an agro-ecological business?

What are the procedures for livestock and agriculture?

Who issues a land title?

What documents do I need to live on a plot of land?

What documents do I need to apply for a permit?

What is the percentage of livestock farming and agriculture in the urban commune of Diégo-Suarez?

What percentage of land is used in the urban commune of Diégo- Suarez?

Where do we go to apply for mining, environmental and construction permits?

Printed by Books on Demand GmbH, Norderstedt / Germany